IRON HULLS

FOR

WESTERN RIVER STEAMBOATS,

BY

THEODORE ALLEN, M. E.

1873.

TRANSVERSE SYSTEM

Length on Water Line	*250 feet*	*Displacement at 5 feet (2240lbs)*	*960 tons*
Breadth of Beam	*38 "*	*Weight of Vessel*	*224 "*
Depth Amidships	*10 "*	*Weight of Machinery*	*134 "*

BUILT 1848

LONGITUDINAI SECTION

MIRZAPORE

SCALE 1"=16'

EXAMPLES OF

SHALLOW RIVER

STEAMERS.

LONGITUDINAL SYSTEM.

Length on Water Line	*198 ft. 3 ins.*	*Displt at 2 ft. Load Draft (2240lbs)*	*331 tons*
Breadth of Beam	*38 "*	*Weight of Iron in hull*	*103 "*
Depth at Side	*6 "*	*Weight of Machinery*	*78 "*

LONGITUDINAL SECTION

J. SCOTT RUSSELL'S PLAN

SCALE 1IN.=16 FT.

AM. PHOTO-LITHOGRAPHIC Co. N.Y. (OSBORNE'S PROCESS)

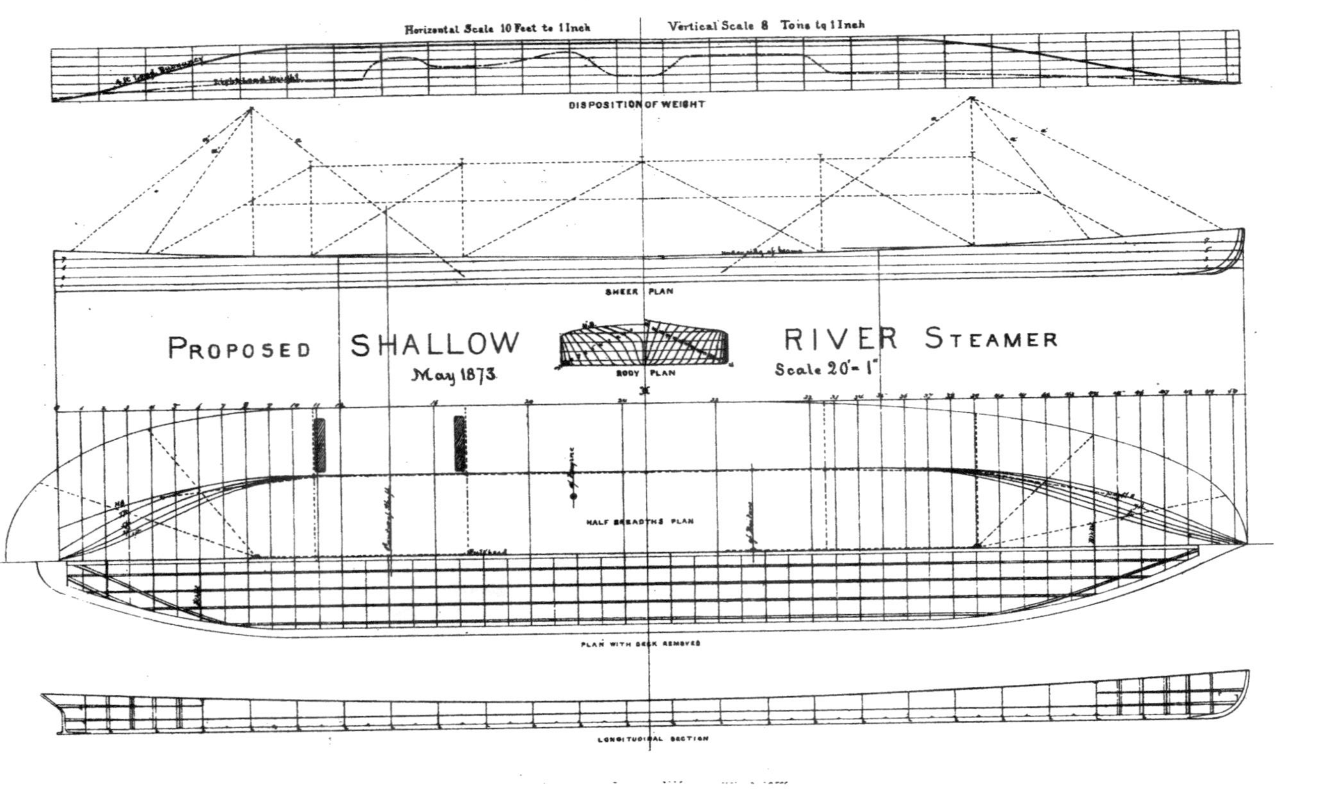
Horizontal Scale 10 Feet to 1 Inch
Vertical Scale 8 Tons to 1 Inch
DISPOSITION OF WEIGHT
SHEER PLAN
PROPOSED SHALLOW RIVER STEAMER
May 1873
BODY PLAN
Scale 20'=1"
HALF BREADTHS PLAN
PLAN WITH DECK REMOVED
LONGITUDINAL SECTION

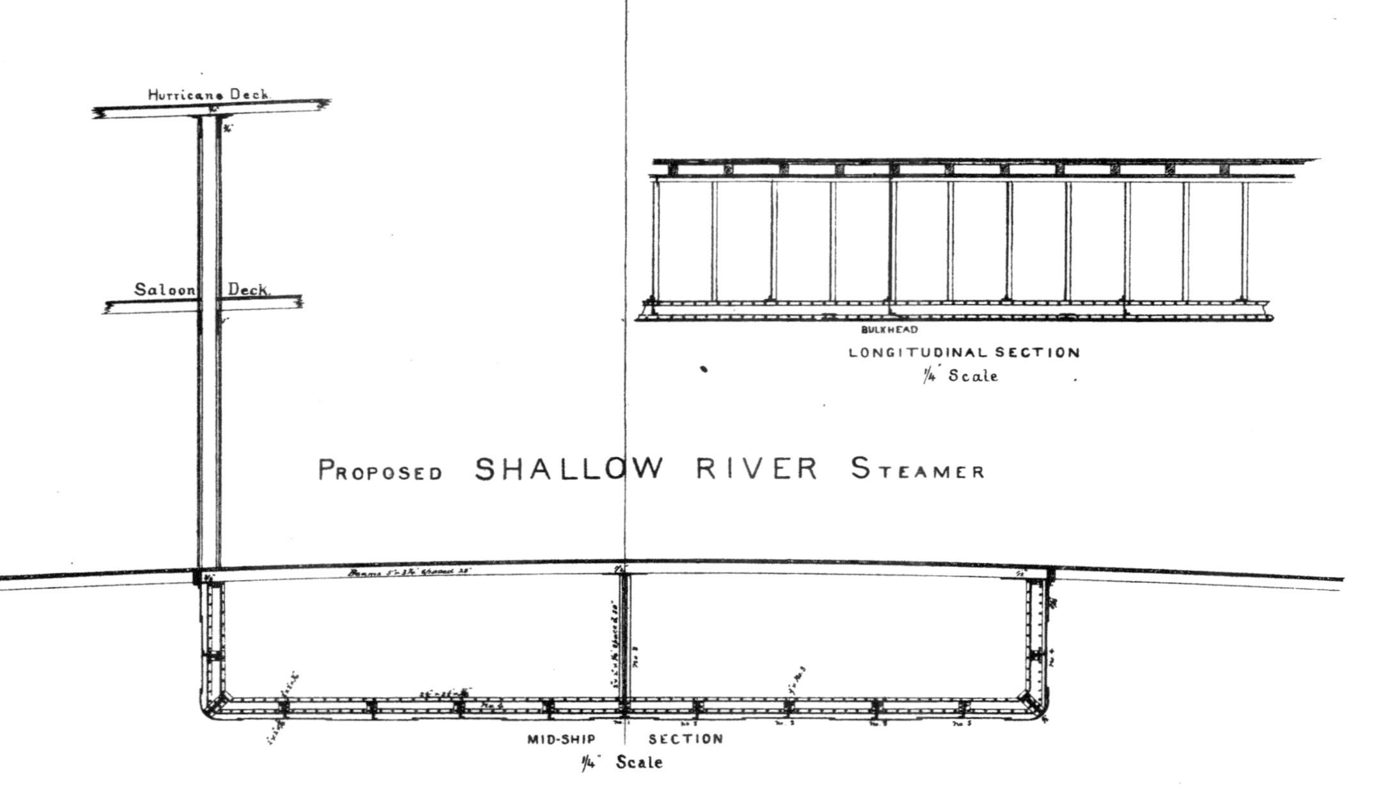

AM. PHOTO-LITHOGRAPHIC Co. N.Y. (OSBORNE'S PROCESS.)

AMERICAN SOCIETY OF CIVIL ENGINEERS.

INCORPORATED 1852.

TRANSACTIONS.

NOTE.—This Society is not responsible, as a body, for the facts and opinions advanced in any of its publications.

LXV.

IRON HULLS FOR WESTERN RIVER STEAMBOATS.

A Paper by THEODORE ALLEN, M. E., Member of the Society,

READ AT THE FIFTH ANNUAL CONVENTION, IN LOUISVILLE, KY., MAY 21ST AND 22D, 1873.

CHARLES HERMANY, C. E., IN THE CHAIR.

Owing to the increasing scarcity of ship timber, and the consequent greater cost of wooden vessels, attention has been more and more turned to the substitution of iron for wood in the structure of hulls for vessels of all descriptions, with steam and sail, for ocean traffic or river service. Abroad, in England especially, iron has almost entirely superseded wood for this purpose, and in this country at the present time out of twenty-one ocean-going steamers in process of construction but four are being built of wood. The greater durability, the lesser weight, the small cost for annual repairs of iron hulls, when compared with wooden hulls, has, even on this side of the Atlantic ocean, overcome the slight increase of their first cost, and the wooden ocean steamer is almost a relic of the past. On the lakes the increase in the number of iron vessels built each year for the past ten years, shows that even in that section of the country where wood should still be cheap, the iron hull is rapidly winning its way to public favor. The use of iron hulls for Western river boats has excited considerable discussion during the past few years, and it is for the purpose of presenting the advantages of the iron hull in more of a professional form than heretofore, that this article has been prepared. To introduce the subject in such a manner as to fairly compare the two methods of construction, it is proposed to give a history of the use and disposition of the material in iron hulls for light draught purposes, up to the latest examples published; to show the various devices which have been used to obtain the great strength, with lightness, necessary for vessels of this class, and to present a plan which, based upon past experience, shall be especially adapted to the requirements of steam vessels designed for service upon the shallow waters of our Western rivers.

The first iron canal-boat of which any record has been preserved, (Grantham's "Iron Ship Building") was built in 1787; she was 70 feet long, 6 feet 8½ inches beam and weighed 8 tons. Some of these early vessels were broken up after 40 years' service; the first iron steamship, the "Aaron Manby," built in 1822, was the first also which went to sea. She was designed for the navigation of the Seine, and carried a cargo direct from London to Paris. Iron river vessels were early built for the Thames, and were soon after introduced in India, upon the river Ganges. On the latter river they were, until 1844, small boats used mainly as tugs, and belonged to the government of India. At that time the Ganges Steam Navigation Company was formed, and from designs prepared by A. Robinson ("Steam on the Ganges") there were five vessels built. They were first put together in England, then taken apart, shipped to Calcutta and there re-erected and riveted up. A short description taken from Robinson's work will give an idea of the progress of iron river constructions up to that time.

The first of these boats, called the "Patna," was launched in 1846. She was intended to carry both passengers and freight, and in model and plan of deck and saloons, was quite similar to the Mississippi steamers. Her dimensions were:

Length on load water-line	195	feet.
Beam over hull	28	"
Beam over paddles	$46\frac{9}{12}$	"
Depth of iron hull	$7\frac{9}{12}$	"
Bottom of hull nearly flat.		
Displacement at 2 feet draft	205	tons.
" " 3 "	328	"
" " 4 "	455	"

The peculiarities of construction were as follows: "There is no external keel; it is replaced by an internal one or keelson, formed of a light, hollow iron beam, 2 feet deep and 9 inches wide, which is riveted to the inner frames of the bottom of the floor. Between this keelson and the iron deck beams, and riveted at their upper and lower sides to both, are light, stiff stanchions of iron, which have the effect of both trusses and ties; binding the floor and deck together. The sides of the vessel are vertical, and the iron frames which run up to form them, finish at the gunwale in a strong cornice, formed of angle iron and a narrow plate. The heads of the frames, the upper edge of the top strake of plate, and the ends of iron deck beams, are thus all riveted together. The powerful connection by this means formed between the bottom and floor and the midship trussing, constitute the entire hull into one large, hollow beam. The sides themselves are for a third of the vessel's length amidships strengthened by diagonal ties, crossing the ribs or frames at an angle of 45 degrees, and riveted to each rib. All the iron in the frame, flooring and shell is of light scantling, but of a quality and make giving the greatest tenacity and strength." "The paddle-boxes are built upon the sides of two light, hollow beams, which cross the vessel under the deck, and project beyond the sides for the purpose; the paddle boxes are framed of angle iron. The entire weight of the vessel, with paddle-boxes, but exclusive of machinery, cabins and stores, is 142 tons." The weight of engines, boilers, propelling machinery and engine bearers is 106 tons; cabins and upper deck 12 tons; water in boilers, coal, stores, &c., 46 tons; making a total weight, with steam up, of 306 tons, on a light draught of 2 feet 10 inches. The speed, by log on the trial, was 11½ miles per hour.

The last two of the five vessels were of greater dimensions, and having been built after the trial of the first vessels, may be taken to represent the improvements suggested by the running of the previous boats. Their general dimensions were:

Length on load water line	250	feet.
Beam on load water-line	38	"
Beam over paddles	66	"
Depth amidships	10	"
Weight of vessel alone, including paddle-boxes and deck houses, but without machinery	224	tons.
Engines, boilers and wheels	134	"
Light draught, without water in boilers	$2\frac{2}{12}$	feet.

"The chief difference between these vessels and the 'Patna' consists in the deck being convex or curved upward transversely, like the back of a violin, and in the bracing or trussing between the deck and the floor of the vessel being in the form of a diagonal lattice-work instead of vertical bars or stanchions, as in the former vessel." "The curvature of the deck was admissible from the absence of cabins and the diagonal framing or spine, from the circumstance of the engines being non-condensing and entirely above deck." As one of these vessels, the "Mirzapore," is considered an excellent example of transverse framing, having great strength for her weight, she is illustrated by a longitudinal section.

All vessels up to this time had been constructed in the same general manner as had been previously followed in the construction of wooden hulls, iron frames being substituted for the wooden ribs, and iron sheathing for the wooden planking. At about this time, however, attention was turned to the strength and stiffness of plate-iron structures. Fairbairn took out his patent for a cellular beam; Stephenson projected the Britannia Bridge, and to test the strength of rolled beams, riveted joints, hollow beams and plate-iron work of similar character, and to determine the relative strength of such constructions, a very thorough series of experiments was tried under the direction of William Fairbairn. The deductions from these experiments developed certain formulas and constants which may be said to have since governed the profession in their calculations upon all structures of wrought-iron.

J. Scott Russell, then engaged in iron ship-building on the Thames, seems about this time to have given attention to the question whether in following out the same method of structure in the iron as in the wooden hull, ship-builders were deriving the best results from the material used. Calculations proved that the hulls of even the best vessels, as then constructed, possessed an excess of strength when considered with reference to the pressure exerted by the water against the sides and bottom, but were deficient in longitudinal strength to resist rupture when supported only at the ends or in the middle; side keelsons and deck stringers were used to obtain strength, but were expensive to construct when intercostal, and occupied valuable room if run upon the top of the floors. In order to obtain this longitudinal strength to a greater degree, with less material, Russell introduced what is known as the "longitudinal system." In this system, as the name implies, the frames, in place of being ribs and running athwartships or transversely, run from bow to stern, are kept rigid by bulkheads and intercostal floors or "partial bulkheads," as he terms them. Increased strength at the bow to resist compression or collapse, in case of collision, and greater stiffness at the stern, are also incidental advantages of this system, especially valuable for river vessels. E. J. Reed, late chief constructor of the British navy, says: "It has been objected to the longitudinal system of framing, that a greater space of unsupported bottom plating is left between the frames than is the case in the vertical (transverse) system. But it has been stated in reply that in case a vessel with transverse frames strikes on a rock, those transverse frames become immediately the most certain agents of destruction to the bottom of the ship, while in the longitudinal system the weakness existing is precisely what is wanted, for it allows the plates between the longitudinals to be indented or even torn through without the general structure of the ship becoming injured." ("Ship-building in Iron and Steel," pg. 95.)

The "Annette" was a vessel especially constructed to show the comparative strength of the two systems with equal weights of material. Russell (in his "Ship-building," pg. 371) gave for the "Annette" longitudinal system an increase in stiffness against sagging or hogging of 5 to 4, or a gain of 25 per cent.

This system of construction has been generally adopted for light draught vessels in England. Russell gives the following description of one built by him (Russell's "Ship-building," pg. 618). "The vessel is about 200 feet long and only 6 feet deep, or about 33 times as long as deep, and to get strength without weight in such a proportion is extremely difficult. She was to carry engines capable of driving her at a speed of 11 to 13 miles on the Thames on 20 inches draught. The general arrangements by which this result was accomplished are as follows: All internal frames were at once abandoned, and the whole ship was built of bulkheads longitudinal and transverse, with longitudinal stringers between and partial bulkheads; in short, the whole structure might be said to be cellular; the difficulty was to get sufficient strength in the center of the ship to carry the weight of the engines and boilers, amounting with water and fuel to 150 tons, this difficulty arising from the extreme shallowness of the boat. I consequently decided that the walls of the cabin on deck should be added to the effectual depth of the ship, and I would consider her as a longitudinal girder, having that depth in the center, and construct her accordingly. I therefore made 105 feet of the middle of her length into a couple of plate girders, 15 feet deep, and these form the sides of the cabins, being perforated with doors wherever convenient; towards the two ends, these longitudinal girders are prolonged under the deck 35 feet at each end beyond the cabin girder, so that effectively these deep girders give strength to the ship through her whole length." The general dimensions of this vessel were as follows :

Length on load water-line..........	198	feet 3 inches.
Breadth over hull..................	38	"
Breadth across paddle-boxes.........	60	"
Depth at side......................	6	"
Draught of water; light............	1	" 8 "
Draught of water; laden............	2	"
Indicated horse-power..............	688	H. P.
Area midship section................	75	sq. feet.
Area load water-line................	6,474	" "
Load, displacement (2,240 lbs.)......	331	tons.
Displacement between 1 ft. 8 ins. & 2 ft.	$61\frac{2}{3}$	"
Grate surface......................	72	sq. feet.
Fire surface.......................	3,300	" "
Weight of iron in hull..............	103	tons.
Weight of engine...................	$40\frac{6}{10}$	"
Weight of boilers..................	$37\frac{4}{10}$	"
Weight of water in boilers..........	22	"
Diameter of paddle-wheels..........	14	feet 4 inches.
Length of buckets..................	9	"
Revolutions of engines per minute...	38	
Pressure of steam..................	25	lbs.

Another steamer, intended for towing on the Rhine, also built by Russell, of about the same length, but having only 25 feet beam, constructed upon the longitudinal plan, had a load draught of 3 feet. She was 9 feet depth of hold, and her longitudinal floors, except under engines, were 16 inches deep only. The engines, boilers and water weighed 128 tons, her displacement at a load draught of 3 feet was 294 tons ; consequently the weight of hull complete, with the coal and the small cargo she may have carried, could not have exceeded 166 tons.

The information I have been able to obtain in regard to American practice is meagre ; so far as ascertained, the boats have been built exclusively on the transverse system, although longitudinal bulkheads and stiffening frames have been used. Haswell (pg. 650) gives the following particulars of a stern-wheel steamer :

Length on deck.....................	110	feet.
Beam over hull.....................	14	"
Beam over guards..................	22	"

Depth of hold.................... 3 feet 6 inches.
Load draught.................... $1\frac{1}{10}$ "
Two cylinders, diameter each....... 10 inches.
Stroke of pistons.................. 3 feet.
Revolutions per minute............. 33

Plating of hull, keel No. 3; bilge No. 4; bottom No. 5; sides Nos. 6 and 7; frames $2\frac{1}{2}\times2\frac{1}{2}\times\frac{1}{4}$ inches, spaced 20 inches; displacement 33 tons at load draught of $1\frac{1}{10}$ feet. Messrs. Harlan, Hollingsworth & Co., have constructed several light draught iron vessels, one a stern-wheel steamer, the "Vengochea," built in 1864, of dimensions as follows:

Length between perpendiculars..... 155 feet.
Beam over hull.................... 26 "
Depth of hold..................... 5 " 6 inches.
Two cylinders, diameter each........ 21 inches.
Stroke of pistons.................. 6 feet.
Mean launching draught, with boilers in........................... $17\frac{3}{4}$ inches.

The Delaware Iron Ship-building Works constructed in 1869 a light-draught steamer called the "Novelty," dimensions as follows:

Length of deck.................... 216 feet.
Beam over hull..................... 24 "
Depth of hold...................... 5 " 6 inches.
Mean draught, with machinery, boilers, water and fuel................. 2 " 1 inch.

The vessels above described will fairly show the present method of construction adopted for river steamers where iron is used. To, however, compare the iron hull with the wooden one, I have taken a wooden vessel of recent construction, combining in its structure and model the latest improvements which experience has shown to be best for the Western rivers; and preserving the same lines and contour, and have replaced the wooden hull with one of iron.

Before entering into the comparison, I will enumerate the primary requirements for a light draught vessel. First—The skin or sheathing must be of as light a material as will withstand the pressure of the water, with the ability to resist the shocks caused by grounding or collision with snags. Second—The frame for stiffening the sheathing and preserving its form must be so disposed as to provide the greatest amount of longitudinal strength with the weight of material used. Third—The bow and bilges must be designed with especial reference to the grounding of the vessel, and the dangers of river navigation due to obstructions.

To meet the first requirements, the strength necessary to resist what may be termed the punching of the bottom, only, will be calculated upon, as the pressure of the water is so slight at the draught sought, that it need not be taken into consideration. To test the relative strength of wood and iron for the skin of a vessel, William Fairbairn, in 1838, tried the following experiments (Fairbairn's "Iron Ship Building," pg. 79): "A plate was fastened upon a frame of cast-iron 1 foot square inside, and 1 foot 6 inches outside. The sides of the plate when hot were twisted around the frame and firmly bolted to it. The contraction by cooling caused it to be very tight, and the force to burst it was applied in the center. This was done in order that the force might in some degree resemble that from a stone or other body, with a blunt end pressing against the side or bottom of a vessel; a bolt of iron terminating in a hemisphere 3 inches in diameter, had thus its rounded end pressed perpendicularly to the plate in the middle." Four experiments were tried with the best Staffordshire iron, two with plates ¼-inch thick, and two with plates ½ inch thick. The results were the plates ¼-inch thick were burst with a mean force of 16,779 pounds; and the ½-inch plates with a mean force of 37,723 pounds. Fairbairn observes "here the strengths are as the depths (nearly), requiring double the weight to produce fracture of a ½-inch plate, as had previously burst the ¼-inch plate." The next experiments were made upon good English oak, of different thicknesses and of the same width as the iron plates; the specimens were laid upon solid planks 12 inches asunder, and by the same apparatus the rounded end of the 3-inch pin was forced through them as follows:

"Mean strength from planks 3 inches thick, 17,933 lbs.

"Mean strength from planks 1½ inches thick, 4,406 lbs.

"Here the strength to resist crushing follows the ratio of the squares of the depths, as is found to be the case in the transverse fracture of rectangular bodies of constant breadth and span."

I wish here to take exception to the general deductions which Fairbairn has drawn from these experiments. The iron tried, from its thinness, possessed no strength when considered as a beam; it was already strained to an unknown extent, from its contraction in cooling. And at best, it was only a test of the tensile strength of the material, when ruptured under unfavorable circumstances. Had the distance between the supports been greater, permitting the iron to stretch and bend to a greater extent, the results obtained would have been greater. The strain being tensile, with similar quality of material, the bursting would of course vary directly in accordance with the thickness. If, on the other hand, the iron had been sufficiently stiff from increased thickness, or by reason of the supports being closer together, so that it could have rested on the supports unsecured, as was the case with the oak, then I believe the resistance to bursting would have been as the squares of the depths. In experiments on the penetration of shot through armor plates, the resistance has been found to follow this general law, which tends to corroborate the view I have taken.

By this experiment it is shown where the distance between the supports is the same, one quarter inch iron plating is equivalent to 3-inch oak planking. If the frames in the iron hull are farther apart, and the iron not previously strained, the iron will, as I have stated, probably stand a much greater relative strain than the experiments showed.

In the vessel taken for comparison, the planking is of 3½-inch oak, the unsupported distance between the frames being 8 to 9 inches; considering the decay of the oak, its water-soaked condition and the narrow planks of which its sheathing is composed, I have upon the above basis estimated that plate-iron of good quality, ¼-inch thick, will resist as great a punching force as 3½-inch oak planking. In regard to the weight; the oak, when water soaked, as it becomes soon after the vessel is launched, will weigh about 5 lbs. a square foot for each inch in thickness, or 17½ lbs. for the 3½-inch planking; while iron of ¼-inch in thickness weighs but 10 lbs. per square foot, and never increases in weight by use. To obtain the full benefit of the yielding of the bottom plating between the longitudinal frames—hereafter described—when the bottom strikes an obstruction, there must be no point from stem to stern between the longitudinals where the elasticity is destroyed by increase in stiffness, due to the securing of bulkheads or transverse intercostal frames to the skin; *that is, no transverse bulkheads or frames must be permitted to be secured to, or even rest upon the skin of the bottom.* As, however, it is necessary that the longitudinal frames or floors shall be stiffened by transverse bracing to keep them up to their proper position, I introduce, every 10 feet in the length of the hull, a partial bulkhead reaching down to within about 3 inches of the skin. The plate forming this partial bulkhead extends 3 inches above the top of the longitudinal floors, and is secured to their topping angle bars by a bar of angle iron, which by thus riding over the floors may be made continuous in one piece from the central bulkhead to the beam shelf, effectually tying all the longitudinals together, and preventing their buckling when under compressive strain. To the bottom of the intercostal plate (3 inches above the skin) to stiffen it, is secured a bar of light angle iron, and the ends of the plate are secured to the longitudinals, by vertical corner pieces. Near each end of the boat a transverse water-tight bulkhead is introduced; the continuity of elasticity is preserved by terminating the plates of this bulkhead at 3 inches from the skin, and tightness is obtained by securing to these plates a strong piece of sheet rubber, or other suitable material, made sufficiently long to permit it to be carried for a short distance along the skin of the vessel, to which it is fastened, as well as to the longitudinal floors, by light strips of wood or iron of sufficient strength to resist the pressure of a head of water equal to the depth of hold. Suppose, then, a vessel strikes a snag just sufficiently near the bottom of the bilge strake to permit the vessel to be forced upon it. It meets first the bilge strake, made, as hereafter described, strong enough to withstand the shock; the vessel raises a little and passes on; the light iron of the bottom between the two longitudinal floors most nearly in the line of the obstruction, is sprung upwards; the vessel perhaps raises a little, the snag itself yields a little, in fact everything gives a little, until the vessel has passed over, or the resistance has become too great for her power, and she stops. Every one is aware how difficult it is to punch a hole through metal when it rests upon a yielding substance; put a piece of iron on a soft pine plank, and attempt to punch a hole through it, and you will realize to some extent the conditions I have just described. A man will hammer away a long time before he can drive his sledge through a large smoke-pipe, while one blow would have sufficed to have ruptured similar iron, had it been so secured as to have prevented its yielding under the impact.

In regard to the second consideration: I have assumed the wooden hull to have been so proportioned as to be possessed of sufficient strength to resist all the strains to which a boat is usually subjected. The estimated area of the oak in the bottom, including stringers and keelsons, is 1,650 square inches, allowing the strength of the oak to be as to wrought-iron as 1 to 5, and on account of the butts being unspliced in the oak sheathing, taking $\frac{1}{3}$ of the remaining area as effective for tensile strength (which is more favorable for the wood than experiments have shown—Fairbairn's "Ship Building," page 75), we have then for area of equal strength for the iron $1{,}650 \div (5 \times 3) = 110$ square inches; there must, however, be added to the aggregate area, on account of deficiency in strength at the butts; as the butts will be double-riveted, 70 per cent. of the total area may be considered as effective, which makes the total area required for the iron 110 ÷ .70 = 157 square inches, to give same strength as the wooden boat had when new. I have, however, for additional strength, increased the area about 20 per cent., and in the proposed design have 190 square inches. The decks are alike in both cases, but I have introduced in the iron hull on each side under the beams a shelf or stringer 24 inches wide by ½ inch thick, and a similar one upon the central bulkhead.

For the purpose of obtaining the greatest longitudinal strength, in which river vessels, from their shallowness, are most deficient, the longitudinal system is used, as giving the greatest strength with least weight. In the design under consideration the floors are 9 inches deep, and the angle-bar frame is fitted close to the skin without sliver pieces, the bar being joggled over the butt-straps. These floors run in the center of each strake of bottom plating, the strakes running parallel to the keel plate, and dying out on the bilge strake; the longitudinal floors ending as bottom frames on the bilge keelson. The butts of the plating in bottom and floors should be planed, and should have double-riveted straps; the plates and bars should be had in as long lengths as possible, and great care should be taken in disposing the butts of plates and bars. The dimensions and arrangements of the iron are shown in the accompanying "midship section." A central bulkhead of plate-iron has been substituted for the wooden bulkhead; it is made water-tight, and securely riveted to the keel-plate, to the transverse frames or partial bulkheads, and to the stringer plate under beams.

In carrying out the third consideration I have made the bilge strake ½ inch thick, and its curved shape and the deep floor or bilge keelson, formed at right angles to a tangent of 45 degrees, give it great strength to resist any force which may be brought against it. At the ends of the boat running to the bulkhead, additional longitudinal stringers are introduced. The stem and stern pieces will be of forged iron, 2 by 5 inches in section, and should run well back on to the keel-plate, which will be thickened to ⅜ of an inch at the bow, and the garboard and next strake will be thickened up at the bilge strake.

The accompanying plans show the general construction of the hull, the distribution of the weight of the hull and machinery, and the buoyancy line on a draught of 4 feet. It will be noticed on one side of the "midship section" plan, and in dotted lines on the sheer plan, I have introduced a truss or hog-frame, for the purpose of making the boat more rigid. This truss is formed by erecting upon wooden struts a hollow, bottomless girder at such height that the hurricane deck will rest upon it. The struts are additionally secured and the ties strengthened by placing immediately under the saloon-deck beams two heavy plates of iron. The position of these girders and ties is such that it is believed they will not interfere with the deck-room or state-rooms. As will be noticed on the plan, the beam-shelf is made lighter when the hog-frame is used. The following calculations, based upon Tate's rule, as adopted by Fairbairn (see his "Ship Building"), will show the gain in strength by the use of this truss. Breaking strain of vessel when supported in the middle—S in this case being taken at 20 tons for tensile strength per square inch; without "hog frames":

$$M = \frac{S}{h} I_{\circ} = \frac{20}{3.6} \times 2{,}710.34 = 1{,}505.7.$$

$$M = \frac{W\,l}{8} \therefore W = \frac{8\,M}{l} = \frac{1{,}505.7 \times 8}{252} = 478 \text{ tons, or } 1{,}070{,}720 \text{ lbs.}$$

With "hog frames":

$$M = \frac{S}{h} I_{\circ} = \frac{20}{17.7} \times 42{,}745 = 48{,}299.44.$$

$$M = \frac{W\,l}{8} \therefore W = \frac{8\,M}{l} = \frac{48{,}299.44 \times 8}{252} = 1{,}533.2 \text{ tons, or } 3{,}436{,}368 \text{ lbs.};$$

or showing an increase in strength of the trussed hull over the untrussed one of more than 3 to 1.

The vessel chosen for comparison of weight is the side-wheel steamboat "Potomac," built for the Ohio and Memphis trade at Cincinnati, Ohio, in 1870. She was of the following general dimensions:

Length on load water-line........................252 feet 6 inches.
Breadth on load water-line..................................36 feet.
Breadth over wheel..66 feet.
Depth of hold...6 feet.
Thickness of bottom planking, oak.........................3½ ins.
Thickness of side planking, oak....................3½ and 2½ ins,
Size of floor timbers forward...3 ins. by 6½ ins.; aft, 3 ins. by 6 ins.
Size of top timbers, oak..........................3¼ ins. by 4 ins.
Distance between center of frames..............10½ ins. to 12 ins.
Has one longitudinal bulkhead on keelson.
Deck beams, oak......5 in. by 3¼ ins.; spaced 28 ins. bet. centers.
Deck planks, pine..................................2 ins. thick.
Draught of water with steam up and no cargo—bow, 2 feet 6 ins.; stern, 2 feet 9 ins.; mean, 2 feet 7½ ins.

Displacement in pounds at 2 feet 7½ ins. draught.....1,100,000 lbs.
Distance from bow to center of wheels..................182 feet.
Distance from bow to center of boilers..................105 feet.
Number of boilers.......................................Five.
Size of boilers....................37½ ins. diameter, 28 feet long.
Grate surface..................................37 square feet.
Number of steam cylinders...............................Two.
Diameter of cylinders...................................24 ins.
Stroke of pistons.......................................7 feet.
Diameter of paddle-wheels...............................32 feet.
Length of paddles.......................................12 feet.
Pressure of steam allowed.....................148 lbs. per sq. in.
Average speed against current..........................10 miles.
Cost of "Potomac," new..................................$85,000.
Usual life, or limit of service of similar boats..........6 to 8 years.

In the accompanying plans the lines, body plan, sheer plan, etc., apply equally to both the iron and wooden hulls; the same deck, and in fact everything above the beam shelf, or upper edge of hull, will be the same in both cases. The dimensions of the plates and angle bars of which the hull is constructed are given in the midship section; the weight of the wooden hull is from actual data. The dimensions of the vessel re-remaining the same, the weight of iron in the iron hull will be as follows:

Plate-iron and butt straps	229,000 lbs.
Angle iron	56,200 "
Rivet heads and points	6,672 "
Forgings	1,636 "
Total weight of iron in hull	293,508 lbs.

The weights of the two boats (the machinery and upper works remaining the same) will then be:

	For the iron hull.	For the wooden hull.
Iron hull	147 tons.	270 tons.
Deck	112 "	112 "
Machinery and wheels	93 "	93 "
Water in boilers	20 "	20 "
Joiner work	40 "	40 "
Fuel, fittings, etc	25 "	25 "
	437 tons.	560 tons.

Mean draught wooden boat as above...............32 ins.
Mean draught iron boat as above.................26 ins.

It must be remembered that the draught of such river vessels should be measured by inches.

Let us assume in round numbers that the iron hull weighs 120 tons less than the wooden hull, and will cost 10 cents a pound, or say $30,000, and we will put the cost of the wooden hull at $15,000 (which is said to be much under its cost). Now, assuming the cost of the wooden vessel complete, at $85,000, the cost of the "Potomac" in 1870, and adding $15,000 for the additional cost of the iron hull, and we have the cost of the iron vessel as $100,000.

The durability of wooden boats on the river, when not destroyed by fire or accident, is variously given as averaging from 6 to 9 years. There is no data upon which to predicate the durability of iron hulls (constantly in fresh water) when well constructed; probably 50 years would not be too great a limit. We will assume, however, that the wooden boat lasts 10 years, and is then sold for $15,000, and that the iron hull lasts 20 years, and is then sold for $5,000 more (on account of the iron in its hull), or $20,000. There must then be allowed an annual depreciation, for the wooden vessel, of ($85,000—$15,000) ÷ 10 = $7,000; and for the iron hull, of ($100,000—$20,000) ÷ 20 = $4,000.

The cost of repairs will be taken as the same in both vessels, and it will be assumed that the vessels are run on equal draughts, say on a mean load draught of 5 feet. The total displacement, in tons of 2,000 lbs., at that draught, is 1,095 tons.

The amount of freight carried on this draught is:

Iron boat..........................	1,095 — 437 = 658	tons.
Wooden boat..........................	1,095 — 560 = 535	"
Greater amount of cargo carried by iron boat on 5 feet draught..........................	123	"

or about 23 per cent. more freight on mean load draught than the wooden boat, which it must be remembered costs no additional fuel, machinery or expense to carry. For comparison, we will assume the gross profits, during a single season, are 25 per cent. on the first cost of the wooden boat during the time of its life; and as the iron boat, without additional expense, carries 23 per cent. more, we will add 23 per cent. of the profit of wooden boat to the iron boat's earnings; we have then:

	Wooden Boat.		Iron Boat.	
First cost..........		$85,000.	......	$100,000.
Gross profit..........	$21,250	......	$26,137	
Annual depreciation.	7,000	......	4,000	
Net profit..........	$14,250	......	$22,137	

This shows for the iron boat a net profit 50 per cent. greater than the net profit of the wooden boat. It is believed that with the same model and equal draughts, less fuel will be consumed in driving the iron boat, as experiments have shown the friction of the water to be less against iron than against wood.

Some doubt may be expressed as to the durability of the iron hull, and it may be thought I have overrated its period of usefulness. It has, however, been often remarked that where iron structures are in constant use, and subject to continual strain, that no perceptible destruction occurs from oxidation; that machinery, the rails of railways, and bridges where much used, do not rust to any extent, although exposed to the weather; and the same fact has been noticed in iron hulls in fresh water. As mentioned before, iron canal boats have been broken up after 40 years' service in England; the hulls even then having been so sound as to cause comment, (Grantham's "Ship Building"). Iron vessels, in this country, especially on our lakes and rivers, have been too recently introduced to permit us to form more than an estimate of their durability. One or two vessels may be cited as examples. The United States side-wheel steamer "Michigan," on Lake Erie, was launched in 1845, and has been ever since in constant service; much of the service has been extremely hazardous, as she has been employed early in the season to open navigation; and late in the season, when new ice was forming, to assist vessels in reaching a harbor. I was informed, in 1870, by a government officer, that up to that time not a dollar had been expended for repairs to her hull. The propeller "Merchant," built at Buffalo by J. C. & E. T. Evans, was launched in 1862, and was one of the first propellers expressly intended for lake service. Her original cost was $80,000, and she was sold last fall, after 10 years' service, for $82,500; the hull, upon careful examination, was found to be perfectly sound. The purchasers, the past winter, added 30 feet to her length, and she is now in service. Other cases might be mentioned, but it is believed that the time is past for prejudice on this point against iron vessels.

It has, I think, been demonstrated that the iron hull, when properly constructed, as compared with an equally good example of wooden hull, possesses the following advantages: the ability, on account of less weight of hull, to carry 20 per cent. more freight upon the same load draught; a durability of at least twice that of the wooden hull; and an increase of more than 50 per cent. in the commercial profit, owing to the causes stated.

I cannot more fitly close this paper than with an extract from Fairbairn's work on "Canal Steam Navigation," published in London in 1831, or more than 40 years ago. "I shall conclude by observing, that the field for improvement in canal and river steam navigation, appears to me most extensive; and if pursued with proper attention to lightness and strength of iron employed in the construction of boats, and the proper disposition of the material, so as to obtain from a given quantity the greatest possible strength, no one can limit the amount of improvement which may thus be obtained in a few years."

Col. Merrill—I am somewhat familiar with the development of iron ship-building in the West, and I find the principal obstacle to its general introduction is the difficulty of persuading Western steamboat men that an iron boat will last sufficiently long to pay for the additional expense. The case is not parallel to that on the great lakes, because, as you are aware, a lake boat is not expected to touch bottom like a river boat; the latter is often stuck on sand bars, and carries heavy spars to push off with.

To the best of my knowledge, there are but two iron steamboats on Western rivers, and both of these were built in Cincinnati. One, the "Alexander Swift," was built at a reported cost of $65,000, while, if constructed of wood, she would not have cost over $40,000. The firm that built her, intended to use her for towing their barges loaded with iron, ore and coal, and of course they threw their profits into the cost of the boat. The other iron boat, the "John T. Moore," was built by the same firm, and sold to private parties, who took her up Red River. I have been informed that during the whole time she has been running, (three or four years) not a cent has been expended for repairs to her hull, while wooden boats in the same trade are constantly on the docks. The Red River is one of the worst for snags in the whole Mississippi Valley, as the existence of the Raft would testify. It will thus be seen that iron boat building in the West is yet in its infancy. Some monitors were built during the war, and a few harbor tugs are now built from time to time, but these are not of the ordinary type of Western steamboats, and are not taken into consideration.

A boat was built at Pittsburgh in 1838, the iron for which, if I am not mistaken, came from England. Probably our manufacturing facilities at that time were not great enough to furnish the plates. When this boat, the "Valley Forge," was launched there was great enthusiasm, and a new era in Western boat building was predicted. But the experiment was not repeated, and the "Valley Forge" went South, and wound up her career in Alabama.

After the war the Government built a number of wooden hull snag boats for cleaning out the snags in Western rivers. These boats may be considered as having two ordinary hulls, connected with each other by a shorter scow placed between them (from midships to the stern), and by a heavy beam at the bows, just above the water surface—the space between the beam and the scow being left open to facilitate the disposal of pieces of snags. The beam is called the butting beam, and its use is to loosen the snag in its bed, the boat being run against it at full speed. After being loosened, the snag is drawn up on the butting beam, and there cut up by steam saws. To give you some idea of the tremendous lifting power required on these boats, I will state that the main hoisting chain is composed of links made out of iron 2 inches in diameter, and yet has often been broken in lifting snags. The chain goes directly from the snag to the hoisting drum. According to Trautwine, its breaking strain should be 84 tons. These boats were originally built with a draught of 4 feet, but they now draw 4½, and on this account do not answer their purpose satisfactorily. They work to greatest advantage in low water, as at that time most of the troublesome snags are visible above water. The boats have also become badly strained by the heavy work performed since they were built.

THE CHAIR—It has not been stated whether there is any destructive oxidation in the iron boat.

COL. MERRILL—We have but little experience on this point; but, to the best of my knowledge, there is no destructive oxidation in fresh water; the greatest wear would come from grounding.

www.ingramcontent.com/pod-product-compliance
Lightning Source LLC
LaVergne TN
LVHW011144110826
845150LV00008B/2498

9781418192938